PERFECT PREDATORS

LIONS ON THE HUNT

Robin Johnson

A Crabtree Forest Book

Crabtree Publishing

crabtreebooks.com

Notes for Readers and Class Discussion

This book is designed to teach readers about core subject areas, build curiosity, and inspire further investigation. Readers are encouraged to build upon what they already know about the subject and engage in topics they want to learn more about. Here are a few guiding questions to prompt reflection and discussion. Possible answers appear in red.

Before Reading

Read the title and look at the table of contents. What do you already know about predators and lions?

- *I know that predators are animals that hunt and eat other animals.*
- *I know that lions are large predators that live mainly in savannas.*

What would you like to learn about lions?

- *I would like to learn about the body parts that make lions such fierce and deadly hunters.*
- *I would like to learn how lions compare to other large cats, such as tigers and leopards.*

During Reading

Pause after reading each page or chapter. What questions do you have about what you read? What are you curious to know?

- *I wonder why lions have become extinct in Europe, northern Africa, and parts of Asia.*
- *I'm curious to know why lions are the only cats that live in groups. I wonder if other apex predators live alone or in groups.*

Make connections with what you are reading. How is this information similar to something you already know?

- *I know that some other predators, such as orcas, hunt and work together as a team to catch prey.*
- *I know that one of the biggest threats to many predators is habitat loss due to human actions, leaving them with fewer places to live and hunt.*

After Reading

Recall key details about the book. What was the author trying to teach readers?

- *The author was trying to teach readers that a lion's strength, speed, and stealth make it an especially ferocious hunter.*
- *The author was trying to teach readers that lion populations are decreasing and that lions are at risk of becoming extinct if they are not protected.*

How did the images and captions help you understand more?

- *The diagrams and labels helped show me how a lion's different body parts help it find and attack prey.*
- *The photographs helped me understand what a lion's savanna habitat looks like.*

Contents

FEAST FOR A KING!

All hail the king of beasts! The mighty lion rules the land with power, strength, and **agility**. It **prowls**, runs, and leaps all over the animal kingdom. The lion's deafening roar tells everyone that this cat is large and in charge. A male lion wears its **mane** like a golden crown. While female lions do not have this thick fur around their necks, they are every bit as powerful as lion kings.

A lion is a perfect **predator** that can take down **prey** animals much bigger than itself, such as this buffalo.

WHAT IS A PERFECT PREDATOR?

Predators are animals that hunt and eat other animals for food. There are many predators in the animal kingdom, but only some are big and strong enough to reign over all others in the **food chain**. These perfect predators are highly skilled hunters with deadly body parts such as paws, claws, and jaws that help them defeat their prey.

CHAPTER 1

STRONG

AS A LION

Lions are fierce **felines** that are built for strength, speed, and **stealth**. Their long, muscular bodies and bone-crushing jaws give them the power to devour most animals.

BIG CATS

Large and Loud

Lions are some of the biggest cats around. They are the second-largest cats on Earth—only tigers are bigger. Male lions grow up to 10 feet (3 m) long and weigh up to 500 pounds (227 kg). Female lions are smaller and lighter than males, but they are still very big—and very scary.

Loud as a Lion

Lions are also very loud. In fact, an adult male lion is louder than any other cat—and most other animals on Earth. Its roar is louder than a motorcycle and can be heard up to 5 miles (8 km) away! Lions roar to communicate with other lions, defend their territories, and show their awesome power.

BENGAL TIGER
up to 10.5 feet (3.2 m) long
weighs up to 650 pounds (295 kg)

LEOPARD
up to 9 feet (2.7 m) long
weighs up to 175 pounds (79 kg)

FOUR THAT ROAR

Lions, tigers, leopards, and jaguars make up a group called big cats. These four felines have large, powerful bodies and are the only cats that can roar.

BUILT TO HUNT

Strength and Speed

Lions have long, muscular bodies that are built for hunting and catching prey. Males and females look different but both are very strong, fast, and **ferocious**.

A male lion's shaggy mane helps protect its throat from other male lions. It also attracts females and makes the lion look bigger and scarier than it is.

Lions have thick golden-yellow fur.

A lion has a long tail with a tuft of black fur on the end. It uses its tail for balance and to communicate with other lions.

A lion has 30 large, sharp, pointed teeth for holding, killing, and tearing prey apart.

Lions have a keen sense of smell and can sniff out prey from far away.

A lion has small, rough bumps on its tongue for scraping meat from bones.

Lions have long, sharp claws that can be pulled back into their powerful paws.

The spots on a cub's fur help hide it from predators. The spots fade as the cub gets bigger and stronger.

A lion has very good hearing. It can hear prey more than 1 mile (1.6 km) away.
Lions have large eyes and excellent vision. Like other cats, they see well even in the dark.
Like all cats, lions have long, stiff whiskers on their faces. They use their whiskers to sense objects and feel their way in the dark.
A lion has four short, strong legs. It can run very fast but only in short bursts of speed. It can also leap long distances.

LION

Lions and tigers are both big cats with muscular bodies, powerful legs, and long tails. But tigers are larger and more aggressive than lions—and much more likely to attack humans.

Lions have shorter, lighter bodies than tigers but are still massive beasts.

Lions do not have stripes, but their golden fur blends in with the dry grasses in their habitats.

Both big cats bite with about the same force—and have some of the most powerful jaws in the animal kingdom.

Lions can swim but do not like to do so. They will usually only take to the water when necessary for hunting or traveling.

VS TIGER

Tigers are the biggest, heaviest, most powerful cats around. Male Siberian tigers—the biggest of them all—can grow up to 12 feet (3.7 m) long and weigh up to 660 pounds (300 kg).

Most tigers have orange fur with bold black stripes. The stripes help tigers hide in the shadows in their forest habitats.

Tigers do not have manes. Adult male lions are the only cats with this furry feature.

Tigers are excellent swimmers and can even hunt from the water.

Both cats have big, sharp teeth that can pierce through animal skulls and bones. A lion's teeth are about 3 inches (8 cm) long—a bit shorter than a tiger's teeth.

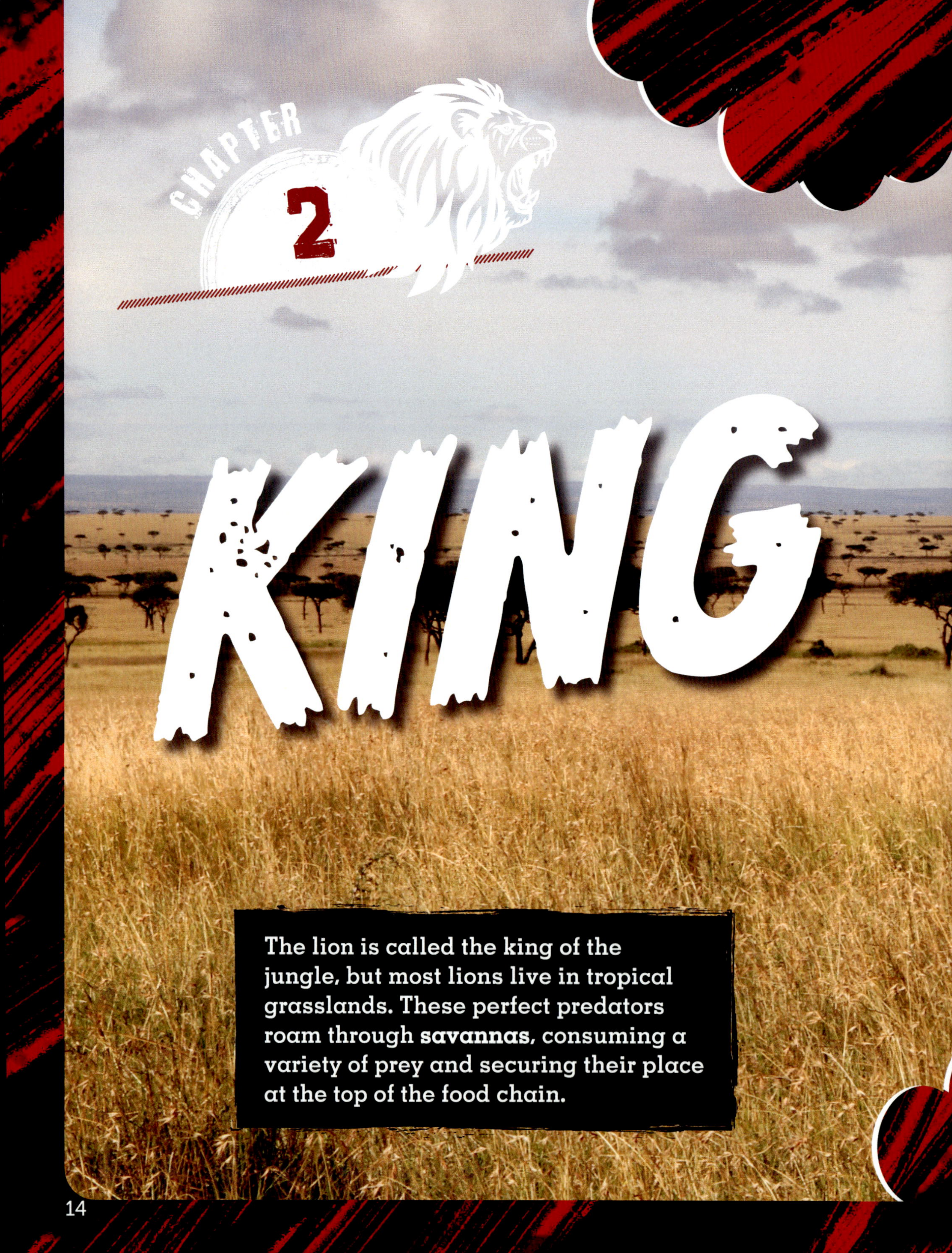

CHAPTER 2

KING

The lion is called the king of the jungle, but most lions live in tropical grasslands. These perfect predators roam through **savannas**, consuming a variety of prey and securing their place at the top of the food chain.

OF THE
SAVANNA

WHERE DO LIONS LIVE?

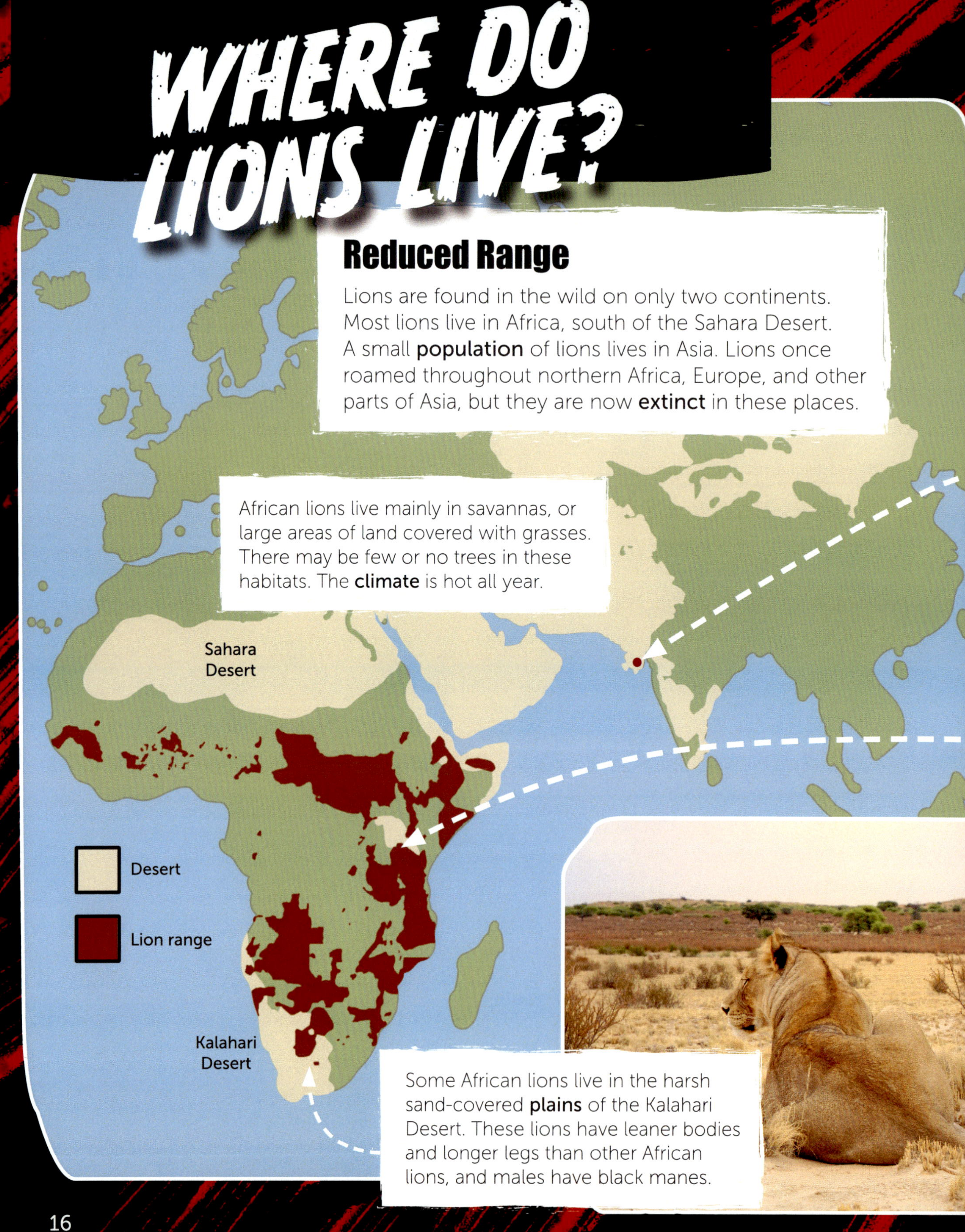

Reduced Range

Lions are found in the wild on only two continents. Most lions live in Africa, south of the Sahara Desert. A small **population** of lions lives in Asia. Lions once roamed throughout northern Africa, Europe, and other parts of Asia, but they are now **extinct** in these places.

African lions live mainly in savannas, or large areas of land covered with grasses. There may be few or no trees in these habitats. The **climate** is hot all year.

Some African lions live in the harsh sand-covered **plains** of the Kalahari Desert. These lions have leaner bodies and longer legs than other African lions, and males have black manes.

Asiatic lions are slightly smaller than African lions, and males have shorter manes.

The entire population of Asiatic lions lives in Gir National Park in western India. The habitat in this protected area is **scrubland** and dry **deciduous** forest.

A large population of African lions lives in Serengeti National Park. About 4,000 lions make their home in this protected area of Tanzania.

CAT HABITAT

Suited for Savannas

Savannas provide the ideal habitats and hunting grounds for lions. A variety of grasses grow there and attract plenty of **herbivores** for lions to eat. The golden grasses provide cover and **camouflage** for lions on the hunt. And the land on the savanna is flat and mostly treeless, which allows lions to spot prey from far away—and then run quickly toward it.

Savanna grasses help hide lions on the hunt.

Cat Naps

Lions spend most of their time sleeping in savannas. They rest for up to 20 hours a day! Lions **conserve** their energy so they can chase and catch large prey such as giraffes. Animals that are lower on the savanna food chain must stay alert and cannot risk sleeping so long. Giraffes, for example, sleep for only 30 minutes a day!

Golden grasses provide camouflage for lions sleeping on the savanna.

Lions love cat naps! They sleep more than any other big cats.

TREES PLEASE

Some lions live in habitats that have trees. The lions climb the trees to sleep or look for prey. They also lie in the shade of the trees to rest and keep cool during the heat of the day.

LION LUNCH

Eat Like a King

Lions are **apex predators**. They are at the top of the food chain and have no natural predators. They eat many prey animals below them in the food chain, which helps keep grassland **ecosystems** in balance. Without lions, there would be too many herbivores that would eat too many plants, and the entire food chain would be **disrupted**.

Lions usually hunt large herbivores, but they will also eat smaller prey animals.

Need for Meat

Lions are carnivores, or animals that eat mainly meat. They usually prey on large animals with hooves, such as zebras, wildebeests, buffalo, gazelles, and young giraffes and elephants. They may also eat smaller animals such as hares, tortoises, and rodents. Lions are **scavengers** too. They eat animals that have died naturally or that have been killed by other predators.

These lions are feeding on the **carcass** of a zebra.

PREDATOR POP QUIZ

Look at the pictures below and match the hungry lion with the foods that make up its main diet.

Answers: A lion eats zebras, wildebeests, giraffes, and tortoises.

LION PRIDE

Strength in Numbers

Lions are social animals that live in family groups called prides. Most prides are made up of about 15 lions, but there may be as many as 40 lions in a pride. Living in prides helps lions stay safe, raise their young, catch large prey animals, and hold their spot at the top of the savanna food chain.

No Boys Allowed

Prides consist of several related female lions, their cubs, and up to three adult males that are not related to other lions in the pride. Young males are forced out of prides when they are two to three years old and must learn to live and hunt on their own.

Lions are the only big cats that live in groups. Most other cats just want to be left alone!

Some male lions live and hunt together.

Band of Brothers

Male lions may form groups called coalitions. Coalitions are made up of a few male lions—usually brothers or cousins—that live and hunt together. When they are adults, the males may fight other male lions to take control of prides.

Other male lions survive on their own in the savanna.

3

BEWARE

Lions are **ambush** predators that use the element of surprise to catch unsuspecting prey. They are highly skilled hunters that often work together to corner, capture, and kill large animals.

STEALTH AND SPEED

Savanna Stalkers

Like all ambush predators, lions must avoid being spotted by the animals they want to catch and eat. They hunt mainly at night or in early morning under the cover of darkness. They have excellent night vision and can spot prey even in very dim light.

Lions often stalk their prey before attacking. They blend in with their surroundings, allowing them to sneak up without being seen.

Crouching and Creeping

Lions crouch in the tall savanna grasses, hiding and waiting for animals to approach. They are smart and patient hunters. They choose places where prey animals usually pass and will wait hours for them to come near. Lions also creep slowly along the ground, silently **stalking** their prey.

Striking Distance

When a prey animal is near, the lion bursts out of hiding and rushes quickly toward it. Lions can run very fast—up to 35 miles per hour (56 kph). But they tire quickly and can only chase animals for short distances. To succeed in the hunt, a lion must always be in striking distance.

Since lions can only run fast for short distances, they wait to charge until their prey is close.

PREDATOR POP QUIZ

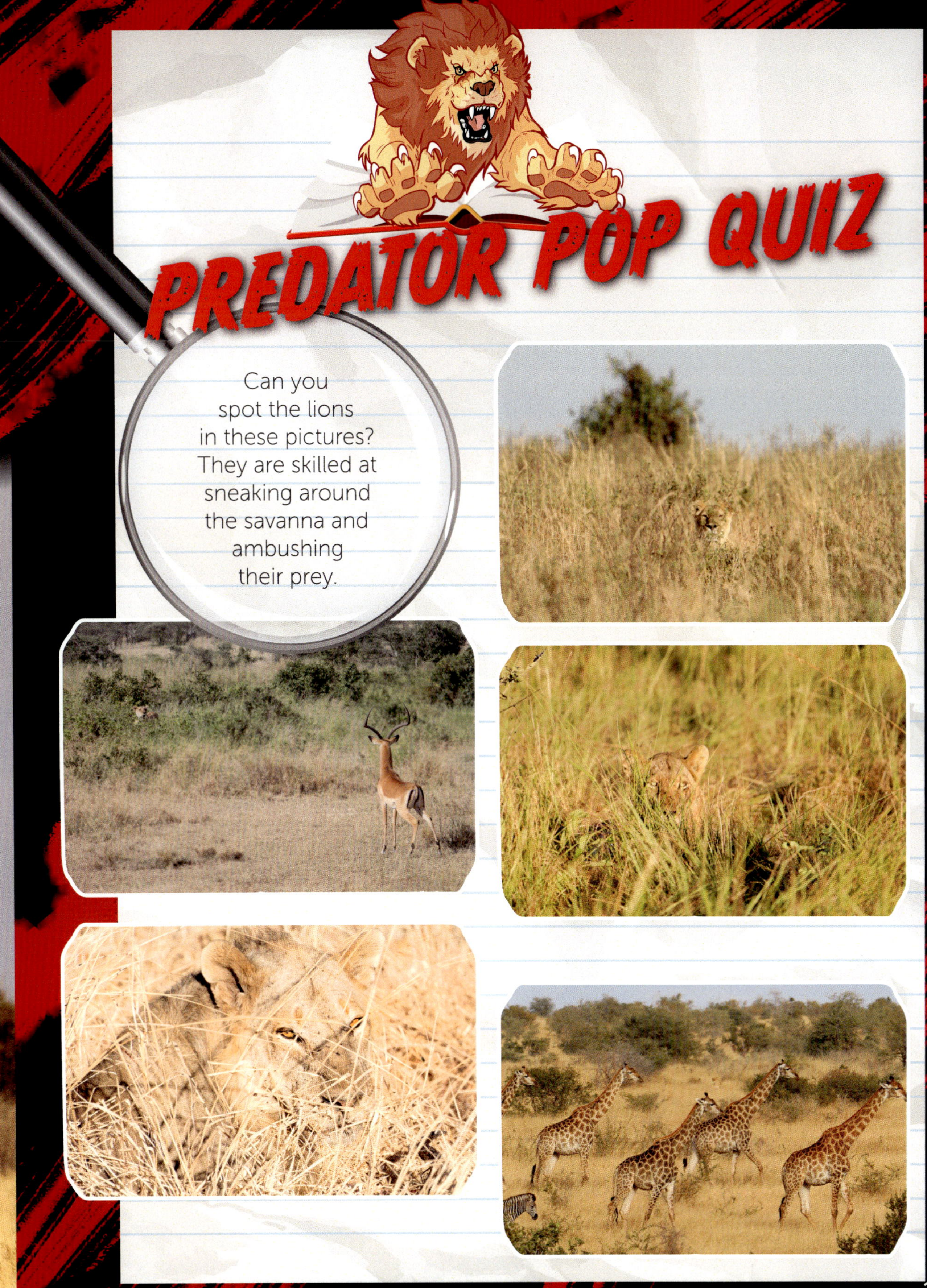

Can you spot the lions in these pictures? They are skilled at sneaking around the savanna and ambushing their prey.

FIERCE FEMALES

Huntresses

All lions are skilled predators, but female lions do most of the hunting in prides. They are smaller and lighter than males, so they are more agile and can run faster. They can also hide better without the large, noticeable manes found on male lions.

Female lions use teamwork to stalk, surround, and chase their prey.

Lions often hunt together, but a single lion makes the kill.

Working Together

Female lions work together to corner and capture prey. They often hunt large herbivores that live in protective groups called herds, so the lions must be smart and stealthy hunters. The lions slowly approach herds from all sides, then attack when the animals start to panic. Smaller lions separate an animal from the herd—usually one that is weak, sick, or young—and surround it.

Making the Kill

The lions pounce on the animal and pull it to the ground. A larger lion then moves in for the kill. She bites the animal's neck and holds on with her sharp teeth until it dies. The lion may also hold her jaws over the animal's mouth and nose to stop it from breathing.

MIGHTY MALES

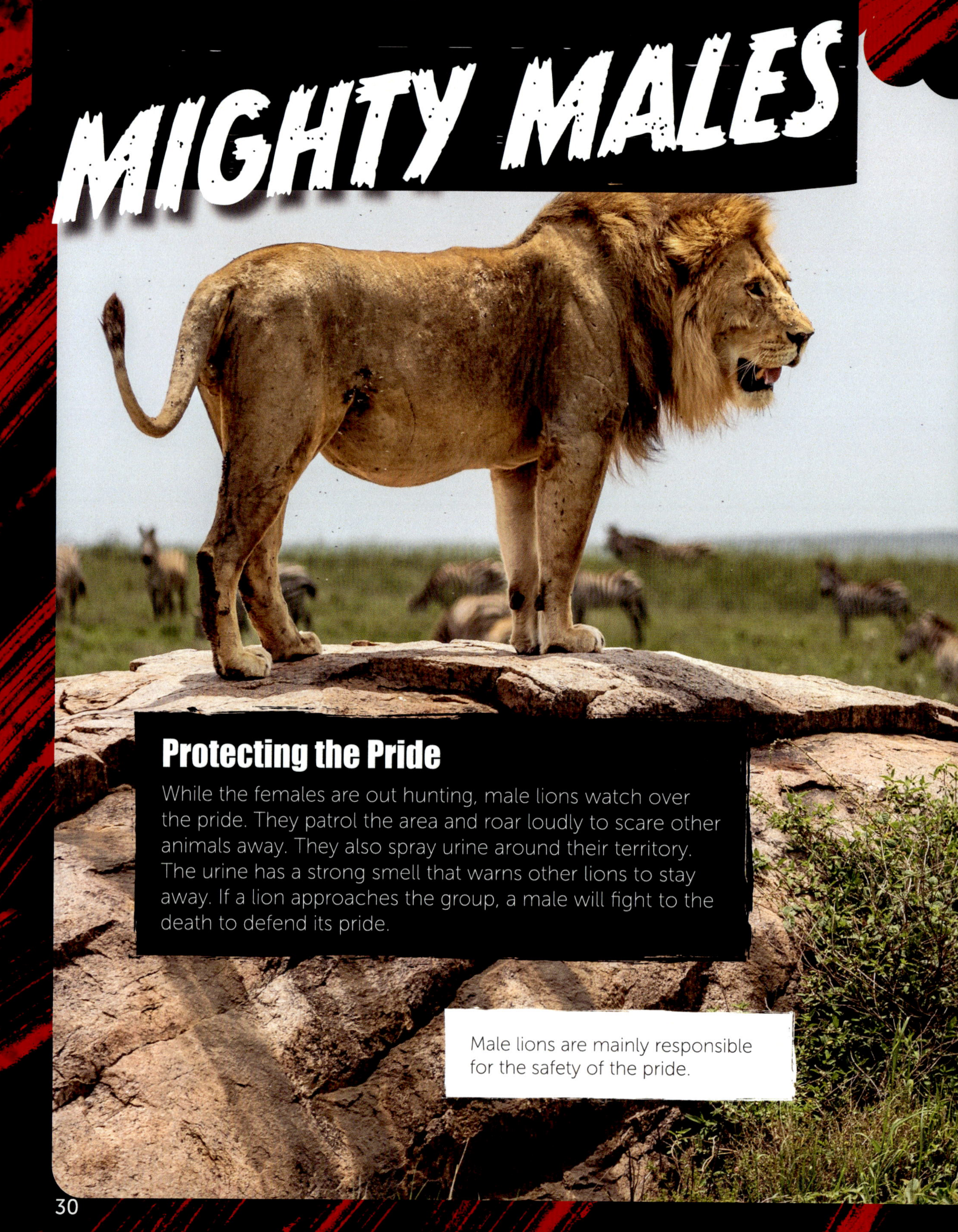

Protecting the Pride

While the females are out hunting, male lions watch over the pride. They patrol the area and roar loudly to scare other animals away. They also spray urine around their territory. The urine has a strong smell that warns other lions to stay away. If a lion approaches the group, a male will fight to the death to defend its pride.

Male lions are mainly responsible for the safety of the pride.

Feasts for Beasts

When female lions finish a kill, the entire pride feasts on it—but the males get the "lion's share". Adult males feed first, ripping off chunks of meat with their teeth and swallowing them whole. Adult females feed next, and finally the cubs get any scraps that are left. Once the pride has finished feasting, scavengers move in to pick the bones clean.

Adult lions eat about 13 pounds (6 kg) of meat each day.

CLEAN-UP CREW

Lions, hyenas, jackals, vultures, and other scavengers play an important role in the savanna. They clean up the grasslands and remove carcasses that otherwise would pile up and spread deadly diseases to animals and humans.

Hyenas, jackals, and vultures move in to finish a carcass left by lions.

A CUB'S LIFE

Not Born Killers

Lions may seem like natural born killers, but cubs are not born ready to hunt. In fact, newborn cubs are small, blind, and completely helpless. They rely on the pride for food and protection.

Caring for Cubs

Mother lions work together to raise the cubs in a pride. They feed and care for their young—and will even **nurse** each other's cubs. The mothers are fiercely protective of the cubs, guarding them from snakes, large birds, and even adult male lions. They teach the cubs how to survive in the savanna and become perfect predators.

Lion cubs play fight to practice their hunting skills.

Female lions teach cubs the important skills of hunting.

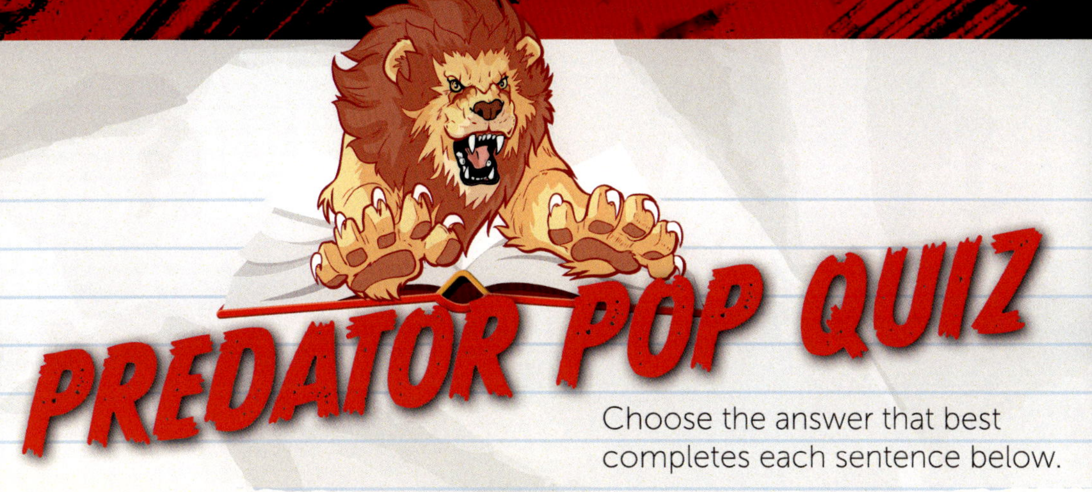

PREDATOR POP QUIZ

Choose the answer that best completes each sentence below.

1 Lions belong to a group called **fat cats / cool cats / big cats.**

2 An adult male lion has long hair around its head called a **mane / main / ponytail.**

3 Most lions live in **Africa / Asia / Europe** in **jungle / wetland / savanna** habitats.

4 Male lions are **louder / bigger / faster** than other cats.

5 Most lions live together in large groups called **teams / prides / herds.**

6 Lions spend most of the day **hunting / eating / resting.**

7 Most of the hunting is done by **male lions / adult female lions / cubs.**

8 Lions are **herbivores / carnivores** that eat mainly **large herbivores / carnivores / scavengers.**

Answers: 1. big cats 2. mane 3. Africa, savanna 4. louder 5. prides 6. resting 7. adult female lions 8. carnivores, large herbivores

LION

Lions and cheetahs are both large, powerful cats, but cheetahs rely on their unmatched speed to hunt and catch prey.

Lions can run about half as fast as cheetahs. Lions tire quickly, so they must plan their attacks and be ready to pounce.

Lions hunt larger prey animals with hooves at night.

A cheetah would probably not win a fight with a lion— unless it cheetah-ed.

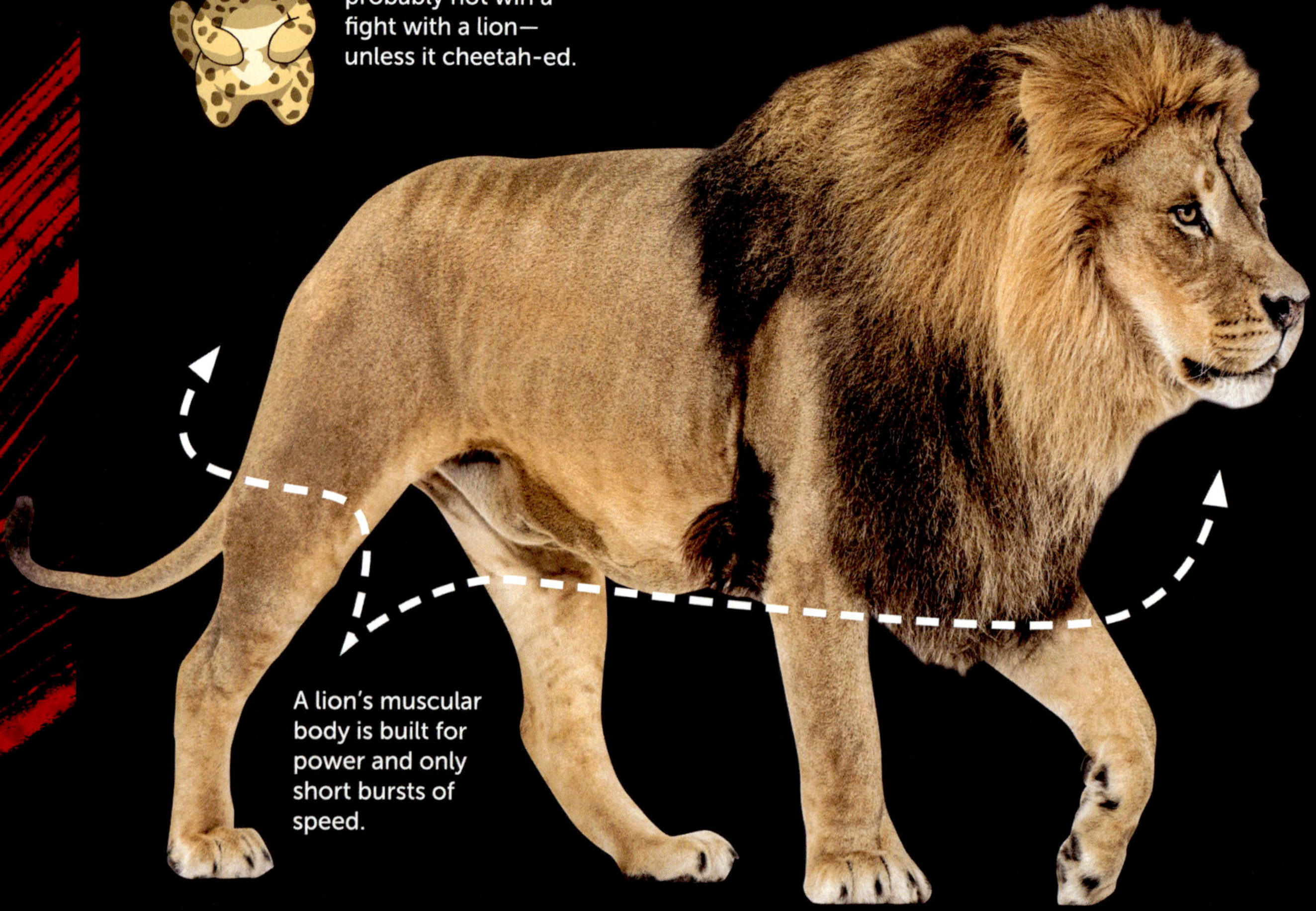

A lion's muscular body is built for power and only short bursts of speed.

VS CHEETAH

The cheetah is the fastest animal on land. It can run up to 70 miles per hour (112 kph). It reaches its top speed in just three seconds and can maintain that speed for about 30 seconds.

Cheetahs have pale yellow fur with black spots. They do not have manes.

Cheetahs cannot roar like lions do, but they chirp, yelp, purr, and make other sounds to communicate.

A cheetah uses its long tail to steer its body and make sharp turns when chasing prey.

A cheetah catches prey by tripping it during the chase. Then the cheetah bites the animal's throat to kill it like lions do.

A cheetah is built for speed. It has a small head, a long, lean body, and powerful legs.

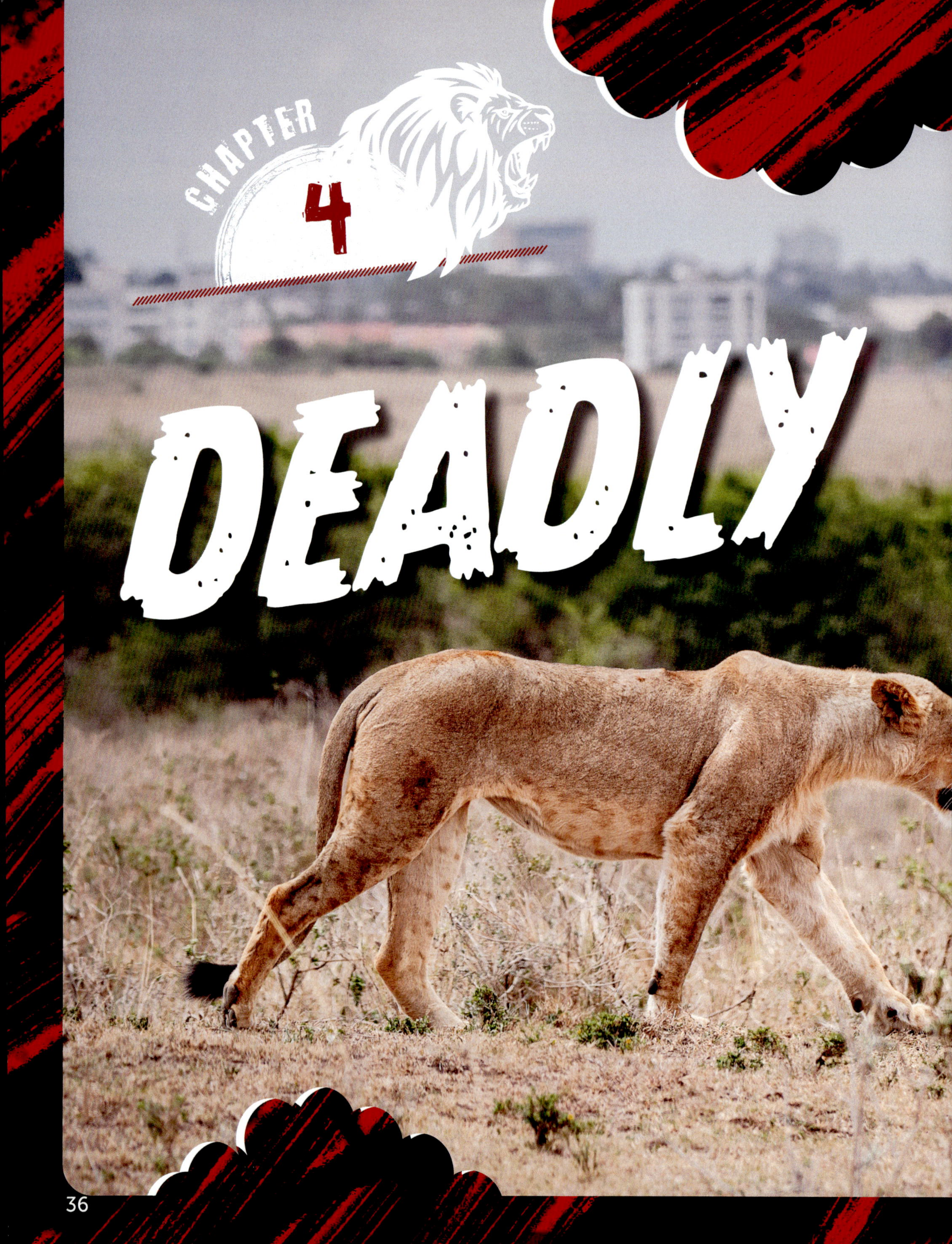
CHAPTER
4
DEADLY

THREATS

Lions are the king of beasts and have no natural predators. Their only real threat is people. Human actions have put lions in danger of becoming extinct, but some people are working to protect these majestic animals.

HABITAT HARM

Shrinking Savannas

One of the biggest threats to lions is habitat loss. People clear the grasses and other plants from savannas so they can build farms, homes, and roads. This reduces the natural places where lions live, rest, and hunt. It also reduces the number of prey animals for lions, since fewer grasses bring fewer herbivores to the area.

Lions may get hurt or killed in wire traps like this one used to catch herbivores.

This kudu was caught in a trap set by hunters. It has a sharp wire snare around its neck.

Lions are losing their habitats as human populations expand.

Bushmeat Trade

People often hunt herbivores for bushmeat in lion habitats. Bushmeat is the meat of wild animals killed for humans to eat. Zebras, buffalo, and many other large herbivores are hunted by people in African savannas, which leaves less prey for lions. Lions may also be injured or killed in the metal traps used to catch herbivores.

Closer to Home

When lions lose their natural habitats and prey animals, they are often forced to live closer to people. Hungry lions may attack and eat cattle and other livestock raised on farms. They do not usually prey on humans, but attacks are becoming more common as lions roam closer to villages.

Lions usually try to avoid living in areas where people live, unless they are forced to because of food shortages or habitat loss.

AT RISK

Killing Lions

People kill hundreds of lions each year. Some farmers kill lions to protect their livestock, pets, and families. They poison carcasses near their farms that they know hungry lions will scavenge. Other people kill lions for their bones, teeth, and claws, which are used to make traditional medicines and luxury goods.

When an adult lion is killed, the entire pride suffers.

Trophy Hunting

People from around the world visit countries in Africa to hunt lions for sport. Trophy hunters pay huge sums of money to stalk lions in the wild and shoot them with rifles or bows and arrows. The hunters are led by local guides and must follow rules set by the government, such as when and where they can hunt. Not everyone follows the rules, however, and many lions are hunted and killed illegally by **poachers**.

Lion furs may be sold in shops like this one in Morocco.

HELPFUL OR HARMFUL?

Many people believe it is cruel to hunt lions for sport. But some argue that it helps lion populations because it controls their numbers and keeps their habitats from being turned into farmland. The money raised by legal hunts provides a source of income for local communities—and a good reason for them to tolerate lions where they live.

Lions are one of the Big Five—five animals prized by trophy hunters as the fiercest and most difficult to hunt. The other prized animals are the elephant, American buffalo, rhino, and leopard.

SAVE THE LIONS

Vulnerable Species

Lion populations are decreasing at an alarming rate. Today there are as few as 23,000 African lions left in the wild. They are considered vulnerable, or at high risk of becoming extinct. Asiatic lions are in even greater danger of dying out. There are only a few hundred of these lions left in India.

Protected Places

To help protect lions and other animals in danger, governments have turned vast areas of land into national parks. Hunting is not allowed in these protected areas—although poachers still hunt lions illegally.

A male lion stands guard over his territory in Chobe National Park in Botswana.

Studying Lions

Scientists observe some lions in the wild. They check the health of the lions and put tracking collars on them so they can follow their movements. Scientists use the data they collect to map the lions' territories, monitor their behaviors, and ensure their populations are stable. They also alert people when lions have roamed outside protected areas.

The tracking collar on this lion allows scientists to monitor her movements.

CECIL THE LION

Cecil was a lion who lived in Hwange National Park in Zimbabwe. He had been studied by scientists for many years and was popular with tourists. In 2015, Cecil was lured out of the protected area with an elephant carcass and killed by an American trophy hunter.

Cecil's death caused worldwide outrage over big-game hunting.

ON SAFARI

Safaris can involve driving in vehicles such as this, walking through the wilderness, or flying in small aircraft.

Looking at Lions

People from around the world visit Africa to learn more about lions. They go on safaris to capture lions and other animals—with their cameras. A safari is an organized trip to find and observe wild animals in their natural habitats, usually in Africa.

At a Cost

Safaris provide local communities with a valuable source of income. People make money running safaris, acting as guides, and working in restaurants and lodges in the area. This reduces the need for poaching and clearing savannas for farms. But safaris also disturb lions in their natural habitats and may cause them stress.

PREDATOR POP QUIZ

Act like a safari guide and explain everything you can about the lions in this picture. What is the group of lions called? Are they males or females? What type of habitat do they live in? Do they usually hunt or rest during the day? What type of prey do they hunt? Add any other information you have learned about lions.

GLOSSARY

agility (uh-JIL-i-tee): The ability to move quickly and easily

ambush (AM-bush): To attack by surprise from a hidden place

apex predators (EY-peks PRED-uh-ters): Predators with no natural predators of their own

camouflage (KAM-uh-flahzh): Colors and patterns that help an animal blend in with its natural surroundings

carcass (KAHR-kuhs): The remains of a dead animal

climate (KLAHY-mit): The usual weather in a place

conserve (kuhn-SURV): To use carefully to avoid waste or loss

deciduous (dih-SIJ-oo-uhs): Describing trees that shed their leaves at certain times of the year

disrupted (dis-RUHPT-ed): Interrupted the normal activity of something

ecosystems (EE-koh-sis-tuhms): Communities of living and nonliving things in their environments

extinct (ik-STINGKT): No longer existing on Earth

felines (FEE-lahyns): Cats or something related to cats

ferocious (fuh-ROH-shuhs): Frighteningly fierce and violent

food chain (food cheyn): A network of living things that shows the order in which animals depend on each other for food

herbivores (HUR-buh-vawrs): Animals that eat mainly plants

mane (meyn): Long hair that grows around the neck and head of male lions

nurse (nurs): To feed a baby or young animal with milk from the mother's body

plains (pleyns): Large areas of flat land with few trees

poachers (pohch-ers): People who hunt and kill wild animals illegally

population (pop-yuh-LEY-shuhn): The total number of a certain species living in an area

predator (PRED-uh-ter): An animal that hunts other animals for food

prey (prey): Animals that are hunted and eaten by other animals

prowls (prouls): Moves around stealthily in search of prey

savannas (suh-VAN-uhs): Broad, grassy plains that are often treeless and found in tropical areas

scavengers (SKAV-in-jers): Animals that feed on dead animals

scrubland (SKRUHB-land): Land that is covered with small trees and bushes

stalking (stawking): Following prey carefully and quietly while hunting

stealth (stelth): Quiet and secretive movements

INDEX

COMPREHENSION QUESTIONS

1. **A small group of male lions that live and hunt together is called a ____________.**
   a) pack
   b) coalition
   c) colony
2. **True or False:** A lion's roar can be heard up to 5 miles (8 km) away.
3. **True or False:** Lions are the largest cats on Earth.
4. **Lions use their ____________ to sense objects and feel their way in the dark.**
   a) paws
   b) tail
   c) whiskers
5. **True or False:** Lions can run fast for very long distances.
6. **Male lions grow up to 10 feet (3 m) long and weigh up to ____________.**
   a) 500 pounds (227 kg)
   b) 600 pounds (272 kg)
   c) 700 pounds (318 kg)
7. **True or False:** Lions often stalk their prey before attacking.
8. **Animals that are at the top of the food chain and that have no natural predators are known as __________ predators.**
   a) supreme
   b) maximum
   c) apex

1. b, 2. True, 3. False, 4. c, 5. False, 6. a, 7. True, 8. c

About the Author

Robin Johnson is a Canadian author and editor who has written more than 100 nonfiction books for children—everything from Animal Celebrities to Food Scientists in Action. When she's not writing, Robin enjoys ambushing her husband and two sons in strategy games and is always on the hunt for the perfect cupcake.

Author: Robin Johnson
Series Concept and Development: Janine Deschenes, Melissa Boyce, Ellen Rodger
Editors: Janine Deschenes and Melissa Boyce
Series and Book Design: Margaret Salter
Proofreader: Ellen Rodger
Print Manager: Candice Campbell
Photographs: Delbars / Shutterstock.com: page 45 (bottom) ; JeanLucIchard / Shutterstock.com: page 40 (bottom inset) ;
Wikimedia Commons: page 39 (top), 43 (bottom)
All other images from Shutterstock.com

Crabtree Publishing

crabtreebooks.com 800-387-7650

In Canada: We acknowledge the financial support of the Government of Canada through the Canada Book Fund for our publishing activities.

Hardcover	978-1-0398-8088-7
Paperback	978-1-0398-8449-6
Ebook (pdf)	978-1-0398-8208-9
Epub	978-1-0398-8328-4

Published in Canada
Crabtree Publishing
616 Welland Avenue
St. Catharines, Ontario
L2M 5V6

Published in the United States
Crabtree Publishing
347 Fifth Avenue
Suite 1402-145
New York, NY 10016

Library and Archives Canada Cataloguing in Publication
Available at the Library and Archives Canada

Library of Congress Cataloging-in-Publication Data
Available at the Library of Congress

Printed in the U.S.A./CP2025